Animals Live Here

Life in the Desert

By Connor Stratton

level 2 little blue readers

www.littlebluehousebooks.com

Little Blue House is distributed by North Star Editions:
sales@northstareditions.com | 888-417-0195

Produced for Little Blue House by Red Line Editorial.

Photographs ©: Pearl-diver/Shutterstock Images, cover; kertu_ee/iStockphoto, 4; the4js/iStockphoto, 7 (top); Isabella Pfenninger/Shutterstock Images, 7 (bottom), 24 (top right); Warren Metcalf/Shutterstock Images, 9; LifeJourneys/iStockphoto, 11 (top); Srabin/iStockphoto, 11 (bottom), 24 (bottom right); alekseiglu/iStockphoto, 12; DavorLovincic/iStockphoto, 15, 24 (top left); WestwindPhoto/iStockphoto, 17; Ingrid Curry/Shutterstock Images, 18; Kristian Bell/Shutterstock Images, 21, 24 (bottom left); Charles T. Peden/Shutterstock Images, 23

Library of Congress Control Number: 2019908248

ISBN
978-1-64619-020-1 (hardcover)
978-1-64619-059-1 (paperback)
978-1-64619-098-0 (ebook pdf)
978-1-64619-137-6 (hosted ebook)

Printed in the United States of America
Mankato, MN
012020

About the Author

Connor Stratton enjoys hiking through deserts, spotting new animals, and writing books for children. He lives in Minnesota.

Table of Contents

Desert Animals

The desert is dry and hot. But many animals live here.

Snakes live in the desert.

They have scales.

Lizards live in the desert.

They have scales too.

snake

lizard

Coyotes live in the desert.

They have fur.

They howl in the evening.

coyote

Hawks live in the desert.

They have sharp claws.

Owls live in the desert.

They have large eyes.

hawk

owl

llama

Big Animals

Llamas live in the desert.

They are big animals.

They have lots of fur.

Camels live in the desert.
They are big animals.
They have humps on
their backs.

hump
camel

Sheep live in the desert.

They are big animals.

They have horns on

their heads.

horn
sheep

rabbit

Small Animals

Rabbits live in the desert.

They are small animals.

They chew on plants.

Hedgehogs live in the desert.
They are small animals.
They have sharp spines.

spine
hedgehog

Hummingbirds live in the desert.

They are small birds.

They have long beaks.

beak
hummingbird

Glossary

camel

lizard

hedgehog

owl

Index